214

i

WHEN COFFEE AND KALE COMPETE
Become Great at Making Products People Will Buy
Alan Klement

Acknowledgments

This book does not represent innovation and insights from one person but
from many. Those listed here—as well as those whose names I have
mistakenly omitted—have helped me understand JTBD and thereby helped
me write this book. I am indebted to them and the entire JTBD community.
Tim Zenderman, Samuel Hulick, Leslie Owensby, Michael Sacca,
Willis Jackson III, Morgan Ranieri, Andrej Balaz, Daphne Lin, Matthew
Woo, Matt Brooks, Mat Budelman, Eric White, David Wu, Bob Moesta,
John Palmer, Rick Pedi, Chris Spiek, Ervin Fowlkes, Timur Kunayev,
Matthew Gunson, Alex Yang, Ryan D. Hatch, Leslie Owensby, Marc
Galbraith, Daryl Choy, James Ramsay, Joshua Porter, Tor L. Bollingmo,
Martin Jordan, Ryan Singer, Laura Roeder, Justin Jackson, Vincent van der
Lubbe, Ash Maurya, Benedict Evans, Esteban Mancuso, Des Traynor, Paul
Adams, Sian Townsend, and the rest of the Intercom team, Daniel
Ritzenthaler, Dan Martell, Anthony Francavilla, Omer Yariv, Justin Sinclair,
Joanna Wiebe, Paulo Peres, Alexander Horré, Bleau Alexandru, Tom
Masiero, Jose A. de Miguel, Dimitri Nassis, Roman Meliška, Paul Gonzalez,
Lee Yanco, Thomas Fröhlich, Lou Franco, David Emmett, Thomas Huetter,
Nir Benita, Kyle Fiedler and Trace Wax and the thoughtbot team, everyone
at the NYC JTBD Meetup, Amrita Chandra, Jeremy Horn, David Lee,
Barry Clark, Ryan Witt, Boris Grinkot, Alex Lumley, Claudio Perrone,
Omar Gonzalez, Ain Tohvri, Amit Vemuri, Sri Vemuri, Hiten Shah, Paul
Sullivan, Matthew Woo, Joanna Wiebe, George White, Dave Rothschild,
Elvin Turner, Mike Rivera, Jason Evanish, Levi Kovacs, and Debbie
Szumylo.
Alan Klement
October 2, 2016
i

Foreword by Rick Pedi and John Palmer

ORIGIN OF CUSTOMER JOBS THEORY

Back in the days of the quality management movement (mid 1980s), the foundational tenet was that customers define quality. And so deploying the Voice of the Customer (the VOC) throughout work processes became the central theme for improving business performance. This was especially true in product development, where design teams obsessed about gathering the VOC. In concept, properly researching the VOC would keep developers on track – not over-investing in features customers did not value, and thus, lose money, while not under-investing in features customers did value, and thus, lose customers. It was from this quality management perspective and the high business hopes of deploying the VOC that we created Customer Job Theory (JTBD) - together with Bob Moesta, Pam Murtaugh, and Julia Wesson.

Because we are the original source of Customer Jobs, Alan Klement has reached out to us to share his JTBD thinking and to ask our point-of-view on the evolution of JTBD Theory in light of its actual foundational roots. With that in mind, he asked us to write this foreword for his book.

BEYOND PRODUCT ATTRIBUTE QUALITY AS VALUE

JTBD Theory represented a major shift in the focus of product developers and the kind of market research used to support product development. The thinking behind Customer Jobs was the hard-learned understanding that developers and marketers needed to adopt a new paradigm about the meaning of value-for-customers. Instead of attaching value to what products are, value should attach to what products do for customers. In other words, stop trying to communicate value with new and improved product features, and start designing more integrated product experiences that are valuable because of what they enable customers to get done in particular contexts of use.

Our thinking called for new market research techniques that got beyond the prevailing methods that had customers evaluating products and describing lifestyle personas. Instead, we grounded customer interviews in recent real-

Foreword

ii

life purchasing and use situations. We learned how customer behavior derives directly from how customers perceive their market-use situation. No one else can know their situations better than customers do. And we would disguise our research purposes to enable customers to tell us what they actually did and why they did it - openly, expressively, and with emotion.

Our early Jobs research work never failed to generate amazing gaps between the reasons producers believed customers purchased products in the category (e.g., products were healthy, indulgent, or more convenient) and actual Job purposes customers had for "hiring a product".

Business growth opportunities could be found in the degree to which the Job that heavy users were hiring the product category to do had not yet been discovered by light or non-users of the category. And optimizing the product design to do the Job better and then communicating the Job value became a good strategy for growing sales.

AN EARLY EXAMPLE

The early JTBD evangelists that we taught and inspired began using Snickers as the de facto standard-bearer for communicating the idea that customers hire products to do Jobs. Ever since 1930, consumers experienced the typical candy bar ingredients of chocolate and nougat combined with whole peanuts more as real food than a candy treat. And yet, it wasn't until 1979 - when Mars introduced the tag "Snickers really satisfies" – that Snickers reflected "new thinking". Much more than a quality cue, "packed with peanuts" gave customers a reason to connect Snickers use with everyday hunger situations.

The Snickers example easily demonstrated several fundamentals of JTBD Theory that are now commonplace:

There are often wide gaps between the value producers think customers assign to their products and the real reasons customers have for using the category

Marketing communications should focus on what a product does for (and to) the customer, not on what it is. A Snickers satisfies hunger – i.e., what it does. It is chocolate and nougat combined with whole peanuts

Foreword

iii

How the design of products can be perfected against the customer's hiring criteria. The food-like qualities of peanuts, Snickers' first bite and chew characteristics, its shape, and its weight in the hand all combine to signal and deliver against the requirements of satisfying hunger in certain customer Job situations

The pitfalls and limitations of defining markets and competition in terms of product categories, versus, seeing markets from the customer's perspective. Mars came to understand that Snickers and Milky Way, which were thought of as "candy bars, chocolate confections, etc." were actually hired by customers for very different reasons. The slogan "Milky Way, comfort in every bar" recognized the differences between customers hiring Milky Way to do a "comfort me" Job and the same customers hiring Snickers to do a "satisfy my hunger" Job.

JTBD THEORY AND INNOVATION SUCCESS

Over the last 15 years, we've continued to advance the thinking we pioneered with JTBD Theory. Our focus has gone beyond using Jobs to explain customer choices. Instead, we have advanced core JTBD principles in order to understand how to develop new product concepts able to carve out new growth footholds in established, fiercely-competitive markets.

We also have followed closely Clayton Christensen's popularization of JTBD Theory and the connections to innovation he and his followers have made. In paraphrasing Clay:

A Job is the progress that an individual is trying to make in a particular circumstance. And for innovators, understanding the Job is to understand what customers care most about in that moment of trying to make progress … i.e., the causal mechanism of customer behavior. Therefore, JTBD Theory provides a way of understanding the foundational question of innovation success: what causes a customer to purchase and use a particular product or service.

Foreword

iv

We respectfully disagree with the assertion that answering that question is the key to innovation success (of course that depends on your definition of innovation). How can you expect to invent new value propositions that will create tomorrow's markets … with an understanding of customer progress that only explains why customers buy products today?

Moreover, understanding what customers care most about in that moment of trying to make progress, brings us back to the very beginning – JTBD Theory as way for businesses to grow sales by improving performance against the customer's definition of quality – essentially aligning today's products and services with the real reasons customers are buying them.

Our work, on the other hand, has lead to asking and answering a very different question as the foundational question of innovation success: how do individuals develop themselves by using the marketplace to advance towards the values, norms, and ideals that enrich the meaning of their lives. After all, innovation is relevant only if it creates or leads customers to substantial new meaning in how they lead their lives. And by focusing on creating new market behavior versus explaining current behavior, we shift the innovator's work from explaining historical choices based on what a product does and is … to understanding why and how customers would develop new market knowledge and behavior tomorrow.

ALAN'S WORK AND WHAT'S NEXT

In light of our work, we find Alan's thinking to be a breath of fresh air. Through determination and perseverance, he has developed the forwardlooking "Self-Betterment" concept to explain the demand side of innovation. He makes a compelling case for Self-Betterment as a basis for innovators to work out what should be next. His case stands in stark contrast to prevailing JTBD Theory that can only explain today's customer choices. Alan's work recognizes that humans have a compelling imperative to improve themselves. His thinking echoes the work we have been doing to explain why people, whose needs are already well met, still hunt the market for new ideas.

In our view, the Self-Betterment idea is on track and Alan is poised to develop it further in this direction: Human beings don't stand still. And in living their lives, they are restless innovators who use their environment - in this case the ecosystem of market information - to imagine better scenarios
Foreword
v
to their status quo … to change existing situations to preferred ones … to imagine what ought to be next in living their lives.

Step-change business growth demands that new product concepts reflect the improvement imperative that keeps customers looking to the market for ideas, devices, and knowledge that break from the past. The "next book" should fill in a picture of work we are now doing with causal models that explain how customers actually use the market to transform themselves and lift their capabilities for living the lives they imagine.
1
1 Challenges, Hope, and Progress
Challenges
Hope
About me
How to be successful with JTBD and this book
Abandon every MBA, all you who enter

This book will help you become great at creating and selling products that people will buy. Your joy at work will grow. You will know how to help companies increase revenue, reduce waste, remain competitive, and make innovation more predictable and profitable. In doing so, you will help economies prosper and help provide stable jobs for employees and the families who depend on them.

I struggled with innovation for many years. I finally made progress when I focused on two things:
The desire every customer has to improve themselves.
How customers imagine their lives being better when they have the right solution.

This understanding has helped me become a better innovator. I believe it will do the same for you. Yet, challenges stand in your way. This chapter introduces these challenges. The rest of this book will equip you with the understanding of how your focus on the customer's desire for self-betterment as a Job to be Done (JTBD) will help you overcome these challenges.
CHALLENGES
Creative destruction is accelerating. The average time a company spends on the S&P 500 continues to drop. In 1960, it was fifty-five years; in 2015, it was about twenty (figure 1). This happened for numerous reasons. A big one is that it has never been easier to create a product and get it to customers.
Challenges, Hope, and Progress
2
This increases the pace at which new innovations disrupt the sales of incumbent ones and then go on to replace them. This process is known as creative destruction.
1
When one innovation wins, another loses. Why? Because a day has only so many minutes, and a customer can use only one product at a time. For example, every day I used to get an espresso from a coffee shop down the street. Two months ago, I bought a Nespresso machine. Now I make my own espressos. The coffee shop has lost my business.
The theory of Customer Jobs and the idea that customers buy a product to complete a Job to be Done, help us understand all the creative destruction

around us. Even though solutions and technologies come and go, the
customer's desire for self-betterment is always there. This focus on
everyone's desire for self-betterment is the key to successful, ongoing
innovation and business.

"Sunk costs" keep us from creating new products. In 1975, a Kodak
engineer invented the digital camera. What was the management's response?
They shelved it. Management argued that Kodak "could" sell a digital
camera, but why would they? They made billions of dollars selling
photographic film. A digital camera would cannibalize their film sales. In the
end, Kodak's management decided that the company would skip digital and
focus on selling photographic film.2

In 2012, Kodak filed for bankruptcy. What happened? Customers no
longer needed film for their cameras–they had switched to digital cameras.
Kodak's downfall was due to management's unwillingness to adapt to a
world with digital cameras–something they had invented forty years earlier.
Very often, it's not legacy technology that stops companies from
adapting, but being tied to a legacy business model. And when change is

FIGURE 1. CREATIVE DESTRUCTION IN ACTION. THE AVERAGE COMPANY LIFE
SPAN ON THE S&P INDEX HAS DECLINED OVER TIME (ROLLING SEVEN-YEAR
AVERAGE).

Challenges, Hope, and Progress
3

proposed to management, they have unlimited excuses to avoid it: "We
make billions of dollars with our current products. Why risk it by selling
something different?"; "We've spent a hundred years perfecting what we do
and building the company we have today. Why should we change?" Excuses
like these make it hard for businesses to adapt, but change will always happen.
Customer Jobs gives you the confidence to break away from legacy business
models and create the products of tomorrow.3

It's a mistake to focus on our customers' physical characteristics. My
father-in-law is sixty-five years old, is from the Bronx, and has never used a
computer in his life. I'm thirty-five years old, from Florida, and wrote my
first computer program at fifteen. Our behaviors, physical characteristics, life
goals, and personal histories couldn't be more different. Nevertheless, we
both own the same model of smartphone. We even use it in almost the same
way. Will a study of who we are and how we behave explain why? Which
data in figure 2 are information? Which are misinformation?4

JTBD helps you become better at knowing the difference between good
data and bad data. This helps you focus on making changes to your product
that bring profits instead of increasing your costs of production only.

We don't take into consideration how customers see competition. In
2006, Indian manufacturer Godrej collaborated with Harvard Business
School professor Dr. Clayton Christensen to create the chotuKool–a lowcost,
feature-minimal refrigerator. It was hailed as a "disruptive innovation"

FIGURE 2. WHAT DATA ABOUT YOUR PRODUCT ARE INFORMATION? WHICH
ARE MISINFORMATION

Challenges, Hope, and Progress
4

that would create a new market of refrigerators and create what Christensen
calls "inclusive growth" for millions of low-income Indians. Unfortunately,
the chotuKool was a costly flop. What happened?

For far too long, academics and analysts–who have no personal
experience with innovation–have created and sold pseudoscientific theories
of innovation. Unfortunately, these theories often mislead. The resulting
product failures exact terrible costs on our economies. This happens because
most, and perhaps all, of these theories don't take into consideration how
customers view competition. Do PCs compete with mainframes because
they're both "computers," or do PCs compete with typewriters, video game
systems, and accountants? Do hard drives compete only with other hard
drives, or do they also compete with tape storage, CDs, DVDs, floppy disks,
flash drives, and cloud storage?

JTBD helps you avoid mistakes like the chotuKool and falling victim to
invalid theories of markets by giving you the knowledge to create an accurate

model of competition before you create a product. It does this by helping you learn how to gain the customer's perspective on what does and doesn't count as competition for a JTBD.

We myopically study and improve on customers' "needs" and expectations of today; instead, we should create new systems that help customers make progress. In the 1860s, the Pony Express was created to help customers get letters and messages across the United States as fast as possible. It lasted only nineteen months. What happened? Western Union established the transcontinental telegraph. While the Pony Express was trying to solve the "needs" associated with using physical mail, Western Union thought, what if we could communicate without using physical messages?5

Very often, innovators think they are studying customers' needs – when in fact they are studying what customers don't like about the products they use today, or what customers currently expect from a product. For many years, manufacturers such as Nokia, Palm, Research in Motion (RIM), and Motorola worked hard to satisfy customers' stated needs and expectations: make a low-price smartphone with a physical keyboard. Today, those expectations have been reversed. Customers don't mind shelling out several hundred dollars for a smartphone, and physical keyboards have almost completely disappeared.

We can't build the products of tomorrow, when we limit ourselves to the needs and expectations associated with the products of today. Instead, we should focus on what never changes for customers: their desire for

Challenges, Hope, and Progress
5

progress. When we focus on delivering customers' progress—instead of what customers say they want—we are free to imagine a world where many needs and expectations are replaced with new ones. Customer Jobs theory helps us ask, "Customers keep asking for a smartphone with a keyboard, but couldn't we help customers so much more if we take it away?" Chapters 13 and 14 show you the power of prioritizing customer progress over everything else.

We may think about the upsides of product changes only, ignore the downsides, and fail to embrace new ways of helping customers make progress. In the early 1980s, the Coca-Cola Company was losing market share to Pepsi. In response, Coca-Cola's management decided to change the formula for Coke, believing that the change would increase market share. They were wrong. Loyal customers went up in arms over it; three months later, Coca-Cola's original formula was restored. Over time, the company regained its market share, but it was lucky. It had the money and resources to recover from the mistake.

Customer Jobs helps you know when it does and doesn't make sense to change your product. Your product might be fine the way it is. Any further investment might increase only your costs of production. Customer Jobs also helps you understand the opportunity costs: what happens when you don't invest in new products, even if it means cannibalizing existing offerings? Kodak knows the cost of not embracing new ways of helping customers evolve: bankruptcy.

Our decision making can be misled when we manage by visible figures only. Customer satisfaction score (CSAT) is a figure that seems straightforward enough. Ask customers to self-report their satisfaction with your product and record their responses. If the CSAT is high, you're doing a good job. Easy, right? Yet, such data and figures are incomplete at best and misinformation at worst. In chapter 14, you'll learn about Spirit Airlines. Customers have consistently rated it as the worst airline in the United States. Nevertheless, it continues to be the fastest-growing and most profitable airline in America. If customers hate it so much, why do they keep flying it? Figures can not only be misleading, they can also be misused. We see this today with the number of monthly active users (MAU) for Twitter. It has experienced explosive growth over the last five years—annual revenue was $106 million in 2011 and $2.2 billion in 2015. Yet analysts and journalists

Challenges, Hope, and Progress
6

continue to write articles titled "The End of Twitter" and "A Eulogy for

Twitter." Why?

The most common criticism is that growth of Twitter's MAUs has stalled at "only" 313 million. Is it any surprise when management's priority then becomes "How can we make MAUs go up?" instead of "How can we continue to make Twitter valuable to users so they won't leave?" Yes, adding new features might push up visible figures such as MAU in the short term, but constant changes might upset and drive away loyal users. Instead, we should congratulate Twitter's employees for their hard work and gently remind them of grandmother's advice: "When you try and please everyone, you end up pleasing no one."6

Many innovators and managers have been influenced by ideas such as:

If you can't measure it, you can't manage it

what gets measured gets improved.

However, such opinions do not take into consideration the following:7

All models are wrong, but some are useful.

FIGURE 3. A TALE OF CREATIVE DESTRUCTION. CHASING VISIBLE FIGURES OFTEN LEADS TO POOR DECISIONS ABOUT YOUR PRODUCT.

Challenges, Hope, and Progress

7

The most important figures are unknown or unknowable, but successful management must nevertheless take account of them.

If you torture the data long enough, they will tell you whatever you want.

It is wrong to suppose that if you can't measure it, you can't manage it – a costly myth.

These statements were made by some of the most important mathematicians and systems thinkers of the twentieth century. They are warnings for those who subscribe to the idea of being driven by visible figures only, and not taking into consideration figures that are unknown or unknowable. Yes, visible figures can be helpful and are often necessary. We have payroll to meet and should strive to increase long-term profits. But we can let figure based data deceive us.

We must remember that data are only proxies for some results of a system. Moreover, the most important figures are unknown and unknowable. What figures or data would have told Apple to remove floppy drives from PCs or keyboards from their smartphones? At the time, many dismissed or criticized these ideas. Journalists claimed Apple's management had lost their minds. Now we regard Apple's decisions as obvious. And what about Twitter's MAUs? The number of users who might want a product like Twitter is a figure that is unknown and unknowable. Twitter's 313 million MAU might represent 100 percent of the market. Analysts, journalists, and even Twitter's own shareholders might be punishing the company even as it achieves market dominance.

There are a variety of consequences that arise when we abandon intuition and risk taking in favor of management by visible figures only. Perhaps the worst are the unfounded beliefs that a product will last forever and that products and companies can continuously grow revenue and attract more customers. The reality is that growth for every product will slow and stop. Nothing lasts forever.

Unfortunately, many managers either don't know or won't accept this. Instead, they become worried when growth slows. They start making changes to their product in hopes of attracting more customers and increasing revenue; however, the effect is often the opposite. Management

Challenges, Hope, and Progress

8

ends up making the product worse for existing customers. With some luck, a competitor won't notice.8

But luck will eventually run out. Another innovation will enter the market with a product that customers find more valuable (figure 3). Why? Because the entrant's innovation cuts off all the baggage the incumbent added during management's frantic attempt to push up all those visible figures. This is when customers begin to switch from the incumbent to the newcomer. So goes the cycle of creative destruction.

Innovation is hard, risky, and nerve-racking. Just ask anyone who has successfully done it. But Customer Jobs can help. With the correct point of view, we can see how visible figures tell us about individual parts of a system only. Once we understand that, we can apply Customer Jobs thinking and understand the relationships around the data. This gives us the ability to assign the proper weight to these figures—or even dismiss them. This helps us become better at knowing if we should continue to improve an existing product, or take a risk and develop a new one.

HOPE

Customer Jobs theory and this book offer you hope whether you are a struggling innovator or just want to become better at understanding, marketing, innovation, product design, or all three. Customer Jobs gives you a collection of principles for understanding why customers buy and use products. This singular attention to customer's desire for self-betterment—instead of what customers say they want, their demographics, or what they do—is what distinguishes Customer Jobs from other theories. This book aims to explain the theory reliably and consistently.

At the time of this writing, no comprehensive book about Customer Jobs theory exists. This is the first. Many others have written interpretations of some of its principles, but almost all of them have created more confusion than clarity.

This book stands out from other writings about Customer Jobs because its contributors—including me—are all innovators and entrepreneurs in our own right. We've applied the theory to our own businesses and products, rather than merely study and preach it.

I developed this book and Customer Jobs theory as if I were creating a product. For Customer Jobs to be successful, it must deliver progress to those who plan on using it. This is why I interviewed sixty-three innovators about Challenges, Hope, and Progress

9

their struggles with innovation and how they designed a product for a customer's Job to be Done. I extracted numerous case studies and insights from these interviews. The most comprehensive and useful ones are featured in this book.

I had to combine my experiences with those of these innovators to design Customer Jobs into what it should be. I learned something from every practitioner I talked with. I am in their debt. Of course, you benefit the most. With this book, you add the accumulated experience of many successful innovators to your own. This collective knowledge will help you become better at creating and selling products that customers will buy.

THE PROGRESS YOU CAN MAKE WITH CUSTOMER JOBS

Customer Jobs is attractive to many because it offers you progress in many ways. Here are a few of my favorite.

Alignment and distributed decision-making. Insights around a customer's Jobs to be Done serves as a "true north". Everyone in the business will use the same customer insights to market, design, build, and manage solutions that customers will buy and use to make progress in their lives. This empowers employees throughout the organization to make good decisions that align with the job, be autonomous and innovative.

Know what data are and are not needed for innovation. Innovators created Customer Jobs theory for themselves. We didn't create it to sell books, collect speaking fees, sell MBA diplomas, or get a PhD from a business school. We created Customer Jobs because we needed help creating successful products that could support our families. In order to do this, we had to figure out what data were and were not relevant to our innovation efforts.

When we design, we're faced the countless trade-offs involved in developing a business strategy, crafting advertising, designing, and engineering, from which questions like these arise:

We can't attack every market. Which ones should we focus on?

Our video ad must connect with customers in just five

seconds. How can we do that?

Challenges, Hope, and Progress

10

Which shade of white will help customers experience luxurious but not sterile?

Which alloy should the suspension be made of to give customers the "feel the road" experience?

Once you know that you need answers to questions like these, you know which data are and are not important. Not only that, you know what should and should not be measured - as well as what can and cannot be measured.

Not everything that can be counted counts, and not everything that counts can be counted.

-William Bruce Cameron

JTBD helps create and sustain a growth culture: what happens when you don't have a healthy growth mindset? You end up wasting time and money building products and features that don't increase revenue. I learned that lesson the hard way - both when I was a Product Manager and as an entrepreneur.

JTBD immunes us with the idea that organizations grow when they offer a growth opportunity to existing and potential customers. No one wants to solve their problems only, we want someone to also give us new and better ways to improve our life-situations in meaningful ways. This thinking helps us answer important growth-related questions such as:

How can we make sure people continue to buy our existing product?

How can we get more people to buy our product, or get them to buy more of it?

What additional products and services can we create that existing customers will also buy?

What additional products and services can we create that can gain us new customers?

Challenges, Hope, and Progress

11

Customer Jobs and its focus on progress gives us a way to think about and answer these questions.

Customer Jobs is a theory evolved over time. The principles of Customer Jobs draw on studies in statistical theory, economics, systems thinking, and psychology. Its principles have slowly evolved over time-at least seventy-five years. It is by no means a flavor of the month.

As a theory, Customer Jobs persists because it is completely decoupled from describing what kind of product you should make. Instead, it is purely focused on understanding how all customers desire to evolve themselves.

You need such a theory to help you become better at innovation.

HOW TO BE SUCCESSFUL WITH CUSTOMER JOBS AND THIS BOOK

I don't believe there is any one "right" way to innovate. Life has too many unknown and unknowable variables. Successful entrepreneur Steve Blank illustrates this by saying, "No business plan survives first contact with a customer."9

Customer Jobs prepares us for whatever curveballs are thrown at us because it equips us with principles instead of methods. Why? Methods come and go, whereas principles stick around. In fact, when you're armed with the right principles, you can plug in any appropriate methods and mental models. You might even create your own methods. And while this idea may cause you some anxiety, over time you will find it empowering. The study and application of principles, instead of methods, is what gives you the confidence to act in a complex world.

As to methods there may be a million and then some, but principles are few. The man who grasps principles can successfully select his own methods. The man who tries methods, ignoring principles, is sure to have trouble.

-Harrington Emerson

While it's easier to teach and learn methods, you become a better designer

and innovator when you ground yourself with principles. This is why this book focuses on principles and demonstrates those principles in action
Challenges, Hope, and Progress
12
through case studies. Most chapters also have a "Put it to work" section that gives ideas on how to apply these principles.

ABOUT ME
I recommend Customer Jobs because I've applied its principles as an entrepreneur, product manager, designer, engineer, and salesperson. I've applied it to my own businesses and on products whose success I was directly responsible for. I believe it's wrong to preach a theory unless you've applied it yourself and were exposed to the risks in the event the theory failed.
Customer Jobs has helped me. I struggled with innovation for many years. I started my first business in 2003. It offered photographic services, image retouching, and a software platform where customers could create their own websites to showcase their artwork. It was a success. At the time, I didn't understand why it was successful. But now I do. It was because I offered a collection of products that worked together—as a system—to help customers make progress. What progress did those customers want? "Help me get recognition for myself and my work."
As my business grew, I saw myself transition from maker to manager. I didn't like it. So, I sold the business and went back to making things. I started another business called Vizipres. It failed. Why? I made a product no one wanted. I lost money, but that didn't bother me. What did bother me was lost time.
I thought that perhaps I needed to learn from other entrepreneurs. I began working for others as an engineer and designer. Later, I began leading innovation efforts at various companies. Inspired by all this knowledge, I began a third business called AIM. It was an advertising marketplace for mortgage brokers and real estate agents. As soon as it got traction and made good money, I sold it to my cofounder. I learned long ago that I'm a maker, not a manager.
That's also when I realized that Customer Jobs, at the time, was incomplete. Many of the principles and suggested theory of the time were inconsistent and often contradictory. I decided to unpack, refine, and expand its principles to help others and myself.
Challenges, Hope, and Progress
13
ABANDON EVERY MBA, ALL YOU WHO ENTER
I invite you to explore JTBD with me. I also ask that you—at least for the time being—put aside any preconceived notions of competition, markets, and even Customer Jobs. You can pick them up again when the book is finished, or you may decide you no longer need them.
14
PART I
THE JOB TO BE DONE
Our journey begins with an introduction to Customer Jobs theory and the idea of a customer having a Job to be Done (JTBD). This section will equip you with a strong foundation to understand what it means to study Customer Jobs and why it is a vital part of innovation.
15
2 What is Customer Jobs? What is a Job to be Done (JTBD)?
My JTBD creates a new me
What isn't a Job to be Done?
A Job to be Done defined
Products enable customers to get a Job Done
Where does Customer Jobs theory come from?

Upgrade your user, not your product.
Don't build better cameras—build better photographers.
—Kathy Sierra
Ten thousand years ago, we were hunter gatherers and used our feet to roam

the earth. Today, we have fast food restaurants and autonomous cars. Why
did we change? Because we have an intrinsic desire to evolve ourselves. We
do this by remaking and adapting to the world around us.
The desire to evolve is in our DNA. It's what makes us human.
Moreover, we do this evolution with purpose. We purposefully use the arts
to evolve ourselves emotionally; the sciences to evolve ourselves
intellectually; and engineering to evolve how we interact with the world.
Purposeful evolution is why we are different from animals:
A bear trying to catch food by the river may think: I wish
fishing could be made better, faster, or easier.
But only a human will think: Fishing is no good. If I could
transform that lagoon over there into a place where I can
breed fish, then I'd never have to go fishing again.
The bear thinks only about what is. Today, it may come up with a better,
faster, or easier way to fish. But tomorrow, it is still a bear that fishes. The
What is Customer Jobs? What is a Job to be Done (JTBD)?
16
human, on the other hand, thinks about what ought to be. Today, she fishes,
but tomorrow that can change. If she could figure out a way to no longer
do the fishing herself, then she can focus on improving herself in other
ways–like building a hut so she could move out of that dank cave.
The bear does not think about evolving itself and its world. It never has
a Job to be Done. The human, on the other hand, does think about evolving
herself. And every time she begins the process of evolving herself, she has a
Job to be Done.
Customers are always beautifully, wonderfully dissatisfied,
even when they report being happy and business is great.
Even when they don't yet know it, customers want
something better, and your desire to delight customers will
drive you to invent on their behalf. No customer ever asked
Amazon to create the Prime membership program, but it
sure turns out they wanted it, and I could give you many
such examples.
–Jeff Bezos
FIGURE 4. WHAT IS REVLON REALLY SELLING?
What is Customer Jobs? What is a Job to be Done (JTBD)?
17
MY JTBD MAKES A NEW ME
Charles Revson, founder of Revlon, perfectly encapsulates a JTBD when he
said:
In the factory, we make cosmetics. In the drugstore, we sell
hope.
With these words, Revson marks the difference between what customers
buy, and why they buy it. This thinking was also carried over into Revlon's
advertising. In 1952, Revlon's breakout advertising campaign was Fire and
Ice (figure 4). The advertising campaign makes it clear: Revlon isn't selling
a product, it's selling a "new me." In fact, there's barely any mention of any
product. One whole page is a check list of provocative questions; the other
features a picture of model Dorian Leigh. Only on further investigation do
you notice the lipstick and nail polish at the bottom of the page.
A Job to be Done is neither found nor spontaneously created. Rather,
it is designed. The checklist of provocative questions such as, "Have you
ever wanted to wear an ankle bracelet?" exists to help customers imagine
(i.e., design) what new me will be created when they buy Revlon's products.
Then there's the picture of Dorian Leigh. Upon seeing that, consumers
continue to design a new version of myself in my mind. For some, the new
me looks like her. For others, the new me is with her. Whatever the case, if
this new me is something I want, I begin desiring it. In other words, I have
a Job to be Done.
All men are designers. All that we do, almost all the time is
design, for design is basic to all human activity. The
planning and patterning of any act towards a desired,
foreseeable end constitutes the design process…Design is

the conscious effort to impose meaningful order.
-Victor Papanek
What is Customer Jobs? What is a Job to be Done (JTBD)?
18
A JOB TO BE DONE DEFINED
Customer Jobs theory states that markets grow, evolve, and renew whenever
customers have a Job to be Done, and then buy a product to complete it (get
the Job Done). This makes a Job to be Done a process: it starts, it runs, and
it ends. The key difference, however, is that a JTBD describes how a
customer changes or wishes to change. We define a JTBD as follows:
A Job to be Done is the process a consumer goes through
whenever she evolves herself through searching for, buying,
and using a product.
It begins when the customer becomes aware of the
possibility to evolve.
It continues as long as the desired progress is sought.
It ends when the consumer realizes new capabilities and
behaves differently, or abandons the idea of evolving.
FIGURE 5. THE DESIGNERS AT INTERCOM (INTERCOM.COM) USE THIS
ILLUSTRATION TO SHOW THE DIFFERENCE BETWEEN WHAT CUSTOMERS BUY,
AND WHAT THEY WANT.
What is Customer Jobs? What is a Job to be Done (JTBD)?
19
PRODUCTS ENABLE CUSTOMERS TO GET A JOB DONE
Humans are limited in our abilities. We can't create a new me by ourselves.
A snap of our fingers cannot create a world where a morning commute is an
enjoyable experience. Realizing such a change requires innovation on the
part of oneself or someone else. Progress can only happen when we attach
and integrate new ideas and new products into our lives. We must become,
as Freud described, "prosthetic Gods."
An example of constructing (i.e., designing) a Job to be Done comes
from a research project I led to understand what Job or Jobs customers were
hoping to get Done (i.e., what new me customers were hoping to create)
with a project management software. Here is a synopsis of one interview.
Notice how the hero of our story comes to realize a new me is possible, and
how he must attach a product to himself to attain that new me:
Andreas began a business around medical tourism. Over
time, he grew his business to include five employees. One
day, he was out with a friend of his, Jamie, at a coffee shop.
During their conversation, Jamie mentioned a product
called Basecamp to Andreas. Andreas had never heard of it.
He was curious to learn more.
FIGURE 6. SAMUEL HULICK USES THIS ILLUSTRATION TO SHOW HOW
CUSTOMERS USE PRODUCTS TO DESIGN A "NEW ME".
What is Customer Jobs? What is a Job to be Done (JTBD)?
20
Jamie explained to Andreas that Basecamp was a project
management tool that helped small businesses become better
at organizing themselves. Andreas was surprised by this. He
knew about complicated project management products like
Microsoft Project, but those were for big companies only,
not smaller ones like his. Currently, Andreas was using
Google Sheets, Google Docs, and e-mail to run his
company. He just assumed that, well, that's how companies
his size operated.
Jamie further explained that Basecamp was made specifically
to help companies his size. As Jamie spoke, Andreas's mind
began racing: Basecamp could help my company stay
organized as it adds more customers and employees. Up
until this point, he had just assumed that his company had
hit its growth limit.
Andreas and Jamie enjoyed their coffee and parted ways.
During his train ride home, Andreas looked up Basecamp

on his mobile device. He also learned about and
investigated similar products to Basecamp. In the end, he
decided to go with Basecamp. He signed up or it, began
using it, and grew his company beyond five employees for
the first time.

This is what a Job to be Done looks like. A consumer goes along his life as
he's come to know it. Then things change. He is presented with an
opportunity for self-betterment—that is, make changes so he can grow.
when or if he finds a product that helps him realize that growth opportunity,
he can evolve to that better version of himself he had imagined. In the case
of Andreas, Basecamp enabled him to gain control over his busisness
could be run. This enabled him to grow it beyond a few employees for the
first time.

Besides demonstrating a JTBD well, Andreas's story also demonstrates
that creating a new me (i.e., having a JTBD) is a process. It's not something
that consumers have; it's something consumers participate in. That's why it's
called a Job to be Done. A comparable example is falling in love. Falling in
love isn't something you have; it's something you participate in. And just as
what is Customer Jobs? what is a Job to be Done (JTBD)?
21
you can't complete the fall-in-love process by yourself, a customer can't
complete a JTBD by himself. He needs a product to help him design,
construct, and complete it.

WHAT ISN'T A JOB TO BE DONE

while many of us have been applying Customer Jobs for a while—Rick Pedi
and John Palmer have been developing Customer Jobs since the 1990s—it
has gained popularity only recently. And like so many things that spread
quickly, many people have distorted and misinterpreted it.

The biggest mistake I see is thinking of a Job to be Done as an activity
or task. Examples include store and retrieve music, listen to music, or make
a quarter-inch hole. These are not Jobs; rather, they are tasks and activities –
which means they describe how you use a product or what you do with it.

For example, music streaming products such as Pandora and Spotify were
designed specifically so customers didn't have to store and retrieve music like
when they used CDs or MP3s. As far as listen to music, that is a broad activity
that varies wildly depending on the context. Someone listening to music so
he can maintain his motivation during a workout is engaging in a very
different activity than someone going to the opera to listen to music.

The problem with describing customer demand as activities is well
articulate by the creator of Activity-Centered Design, Don Norman:
Harvard professor Theodore Levitt once pointed out,
"People don't want to buy a quarter-inch drill. They want a
quarter-inch hole!" Levitt's example of the drill implying
that the goal is really a hole is only partially correct,
however. when people go to a store to buy a drill, that is
not their real goal. But why would anyone want a quarterinch
hole? Clearly that is an intermediate goal. Perhaps they
wanted to hang shelves on the wall. Levitt stopped too
soon.

Once you realize that they don't really want the drill, you
realize that perhaps they don't really want the hole, either:
they want to install their bookshelves. why not develop
what is Customer Jobs? what is a Job to be Done (JTBD)?
22
methods that don't require holes? Or perhaps books that
don't require bookshelves. (Yes, I know: electronic books,
e-books.)

Besides, there are already brilliant design methods out there to help you
design for tasks and activities. Examples include Activity Theory, Don
Norman's Activity-Centered Design (figure 7), Cognitive Task Analysis, and
Human-Computer Interaction (HCI). If you want to learn more about how
to design for activities, go there.

There are not different types of Jobs. Another common mistake is to think that there are types of Jobs. In particular, some may think there are emotional, function, and social Jobs. I'll describe why it's a bad idea from both a practical and theoretical perspective.

Practically, you'll be more successful when you think of every Customer Job as unique. We've learned that while many Jobs share the same core emotional desires (e.g., belonging, self-expression, control, etc.), each Job is a unique combination of these core desires. That is why a good product should deliver on these core emotional desires in its own way. A good example is Facebook. A lot of people use Facebook because it taps into desires such as control, self-expression, and belonging—but it does so in its own unique way. So instead of saying that there are types of Jobs, you'll be much better off thinking that each Job is unique.

Theoretically—that is, from an ontological and epistemological perspective—Customer Jobs are design (artificial) problems, not natural problems. Natural problems are falsifiable. This means they can be objectively measured and determined as either true or false:

FIGURE 7. DON NORMAN'S 1988 BOOK, THE DESIGN OF EVERYDAY THINGS, FEATURES ACTIVITY-CENTERED DESIGN AND THE SEVEN STAGES OF ACTION SEEN HERE. THIS THEORY FORMED THE BASIS OF METHODS SUCH AS TASK ANALYSIS AND HUMAN-COMPUTER INTERACTION.

What is Customer Jobs? What is a Job to be Done (JTBD)?

23

Q: Is argon (Ar) a noble gas?
A: If under conditions X it reacts, then yes; otherwise, no.

Design problems, on the other hand, are not falsifiable and cannot be measured objectively:

Q: Is this painting any good?
Person 1: "Yes."
Person 2: "No."

With respect to Jobs, then, no objective test can be created to say, "This is a social Job. That is not a social Job." If I buy a Ferrari to impress other people, is it a "social" Job because I reference other people? Or should we rephrase it as an insecurity, making it a "personal" or "emotional" Job?

And because there's no way to objectively define each type of Job, every person on the team will have his or her own opinion of what type of Job it

FIGURE 8. ARE YOU DESCRIBING A JOB TO BE DONE, OR SOMETHING ELSE?

What is Customer Jobs? What is a Job to be Done (JTBD)?

24

should be. Moreover, even if/when you do get consensus, so what? Isn't knowing that I bought a Ferrari because I want to "fit in" good enough? What do you gain by labeling it a social or personal Job? I'll tell you: absolutely nothing.

Take it from me, don't waste your time trying to dissect Jobs into different types. It's about as productive as trying to answer, "How many angels can dance on the head of a pin?"

Is it a Customer Job? Does it describe a "new" me or something else? When presented with a possible description of a Customer Job, the best framework of thinking I can offer you is the decision tree in figure 8.

Keep in mind that a Job to be Done describe the "better me." It answers the question, "How are you better since you started using [product]?" Renowned psychologist Albert Bandura described humans as "proactive, aspiring organisms". Customer Jobs carries this idea into markets, making the claim that we buy and use things to improve ourselves, to make progress. If you're not describing a Customer Job in terms of progress, you're probably describing something else.

People are proactive, aspiring organisms.
-Albert Bandura

WHERE DOES CUSTOMER JOBS THEORY COME FROM?

The greatest—and most helpful—theories are not created by one person but are the result of many people over a long period (figure 9). This is certainly the case with JTBD. Its principles have emerged from the work of a long lineage of researchers and innovators. Here are the most notable.

Joseph Schumpeter and creative destruction. The roots of JTBD go back at least seventy-five years to Joseph Schumpeter and his introduction of creative destruction. Schumpeter observed that new innovations steal customers from incumbent offerings and then eventually go on to replace them. At one time, horses and ships were our primary methods of personal transportation. Eventually, trains replaced horses, but then cars and airplanes replaced those trains and ships.

What is Customer Jobs? What is a Job to be Done (JTBD)?

25

Customer Jobs incorporates Schumpeter's insights as it seeks to understand why customers pick one way of doing things over another. Yes, innovators create new solutions, but the wheels of creative destruction turn only through the interaction between customers and innovators.

Customer Jobs also incorporates another one of Schumpeter's brilliant insights that is almost always overlooked. Schumpeter argued that competition should not be measured only among products of the same "type." He insisted that competition can come from anywhere. You might think you're alone in a market or have market superiority, but some competitor unknown to you could be stealing away your customers. Your only sign that something is wrong is decreasing sales. In chapter 8, we take a close look at Customer Jobs, creative destruction, and competition.

W. Edwards Deming and systems thinking. Schumpeter's influence on Customer Jobs is restricted mostly to factors of market dynamics and competition; however, W. Edwards Deming has influenced Customer Jobs the most. Those who are familiar with his nearly sixty years of contribution to theories of management and innovation will recognize his fingerprints throughout this book.

Deming's most notable influence comes from his development of systems thinking, which I discuss in chapter 13. Throughout Deming's

FIGURE 9. A GENEALOGY OF CUSTOMER JOBS.

What is Customer Jobs? What is a Job to be Done (JTBD)?

26

career, he frequently reminded businesses that producers and customers are connected by systems:

The customer and producer must work together as a system.

The consumer is the most important part of the production line.

Deming often challenged companies to remember creative destruction. He impressed on business leadership that simply making a product better and better—improving what already exists—wasn't enough. Sooner or later, someone will invent something new. He would tell businesses the following:

Makers of vacuum tubes improved year by year the power of vacuum tubes. Customers were happy. But then transistor radios came along. Happy customers of vacuum tubes deserted vacuum tubes and ran for the pocket radio. A dissatisfied customer does not complain; he just switches.

Deming understood that improving products of today continually isn't enough: "We must keep asking, what new product or service would help our customers more? What will we be making five years from now? Ten years from now?" For Deming, the process of innovation should never stop.[11]

Psychology. On the psychology front, you'll run into influences from Gary Klein, Amos Tversky, Daniel Kahneman, George Loewenstein, and Ann Graybiel. These are psychologists and scientists whose work forms the foundations of behavioral economics and naturalistic decision making (NDM). Their work helps us understand how and why customers don't make rational decisions when buying and using products, are inconsistent in their opinions of products, and don't always act in their best interest. Customer Jobs understands that if you want to make a great product and to develop a message that connects with customers, you have to understand the emotional forces that shape their motivation.

What is Customer Jobs? What is a Job to be Done (JTBD)?

27

Bringing it all together. Then, you arrive at John B. Palmer, Rick Pedi, Bob
Moesta, Julia Wesson, and Pam Murtaugh. In the 1990s, they began working
together to combine their respective experiences into the first Customer Jobs
principles. They are the ones who came up with the idea and language that
customers have "Jobs" that they are trying to get "Done."
Then, you get to me and this book. John, Rick, and Bob have
personally influenced me more than I could ever express. Last but certainly
not least, the entire JTBD community has had a tremendous influence on
me. Without their application of and experience with JTBD, this book
would not have been possible.
28
3 What Are the Principles of Customer Jobs?

Customer Jobs principles
Here are a few JTBD principles that you will see demonstrated repeatedly
throughout this book. There are others, but the principles below are perhaps
the most useful and commonly used in the Customer Jobs community.
PRINCIPLES OF CUSTOMER JOBS
Customers don't want your product or what it does; they want help making
themselves better (i.e., they want to evolve, make progress). Charles Revson
knew that customers didn't want cosmetics, which are just colored oils. He
also knew that customers didn't want what those cosmetics do, which is
simply coloring skin. He understood that his customers wanted hope. This
understanding of customer motivation has helped keep Revlon in business
for eighty-four years. In 2015, its revenue topped $1.9 billion. It seems that
selling hope is a profitable business.17
Focusing on the product itself, what it does, or how customers use it
closes your mind to innovation opportunities. For example, if you sold drills,
you might be tempted to think that people buy drills and bits because they
want holes. But then 3M comes along and develops an entire line of damagefree
hanging products that are designed specifically to eliminate the need for
a drill or for making any holes. Another manufacturer, Erard, also avoided
the "customers want holes" trap. It promotes a collection of TV mounts
with a simple description: "The first TV wall-mount bracket with no drilling
of the wall required." While you were convinced customers wanted holes,
your competitors understood that customers wanted help improving their
lives.18
People have Jobs; things don't. It doesn't make sense to ask, "What Job
is your product doing?" or say, "The Job of the phone is…" or "The Job of
the watch is…" Phones, watches, and dry-cleaning services don't have Jobs.
They are examples of solutions for Job.
Products don't have lives to make better. They also don't have
motivations, aspirations, or struggles. However, people do struggle. They do
have lives they want to improve. This is why people—not products—have
a JTBD.
What Are the Principles of Customer Jobs?
29
Competition is defined in the minds of customers, and they use progress as
their criterion. Imagine an entrepreneur who wants to be advised and
inspired by someone whom she respects. She has a variety of options to
choose from to achieve this. Examples include reading books, watching
videos, attending conferences, or giving advisory shares in exchange for
mentorship.
The struggling entrepreneur cares little about how she gets advised and
inspired. The concern is about making progress. "Are things better for me
today than yesterday? Am I getting closer to that picture in my mind of how
I want my life to be?" These are some of the criteria customers use to judge
which products compete against one another to improve themselves.
Customers don't define or restrict competition based on the functionality or
physical appearance of a product. Instead, they use whatever helps them
make progress against a JTBD.
When customers start using a solution for a JTBD, they stop using
something else. Many solo entrepreneurs struggle with feelings of isolation,

and hope to be motivated and inspired. To get this Job Done, some choose
to create local get-togethers through Meetup.com and encourage other solo
entrepreneurs to join. If that doesn't work, they may try getting people
together in an online chat group. If that doesn't work, they may decide to
join an existing online community, such as Product People Club. If Product
People Club, as a product, is something that works to make their lives better,
they stop searching for new solutions. Their Job is Done.

These entrepreneurs were jumping from one solution to another. This
makes competition for a JTBD a zero-sum game. For somebody to win,
somebody else has got to lose. Just as only one puzzle piece can fit into an
empty slot, a customer prefers only one solution at a time for a JTBD.

Innovation opportunities exist when customers exhibit compensatory
behaviors. Baking soda was originally advertised as a baking agent. Over
time, customers started using it as a cleaner and deodorizer. Arm & Hammer
picked up on this. It now sells a variety of baking soda-based products
specialized for various cleaning and deodorizing purposes.19

The Segway was meant to revolutionize personal transportation for the
masses. It failed; however, it did find success among members of law
enforcement who began using it for their patrols. Tour companies also began
using Segways as the ultimate gimmick to attract tourists and for family
activities.20

What Are the Principles of Customer Jobs?

30

Baking soda and the Segway are examples of customers using products in
ways for which they weren't originally intended. Such situations represent
opportunities to innovate a new product or to refit an existing one.

Solutions come and go, while Jobs stay largely the same. Customer Jobs
is about understanding our intrinsic desire to evolve. This motivation
changes slowly. Therefore, Jobs change slowly. How long have people
wanted to be mentored and advised by someone they admire? How long
have parents wanted to teach their children life lessons? The answer is the
same: a long time.

Products, on the other hand, constantly change because technology
enables better ways of creating solutions that solve our Jobs. This is why we
focus on the JTBD and not the product itself or what the product does.

Favor progress over outcomes and goals. Customer goals and outcomes
are only the results of an action. The ball went into the net; that is a goal.
Did you win the game? Are you becoming better at making goals? No one
knows.

Measure progress instead. Making a goal today isn't as important as
becoming better at making goals in the future. This philosophy is the same
for your customers. They don't wait until after they've finished using a
product to determine whether they like it. They measure progress along the
way. Do people wait until they lose ten pounds before judging whether a
gym membership is successful?

Customers need to feel successful at every touch point between
themselves and your business, not just at the very end when the outcome of
an action is realized. Design your product to deliver customers an ongoing
feeling of progress. Over time, you will notice that you need to change the
outcomes and goals you deliver to customers. Why? A successful product
and business will continually improve customers' lives. As customers use
your product to make their lives better, they will face new challenges and
desire new goals and outcomes.

Progress defines value; contrast reveals value. See how easily you can answer
this question: "which food do you most prefer: steak or pizza?" Many
people find this difficult to answer. An easier question might be, "when do
you prefer steak, and when do you prefer pizza?"21

A customer may find it difficult to compare two foods without any
context. The last question is easier because the person being asked is thinking
about food and context together.

What Are the Principles of Customer Jobs?

31

Products have no value in and of themselves. They have value only when customers use them to make progress. The value of steak is easier to assess when it's matched with a fancy restaurant and a nice bottle of wine. But things can get wacky in that scenario if we swap a slice of pizza for the steak. The same effect, of course, also applies to pizza. A pizza birthday party for an eight-year-old makes perfect sense, but a steak birthday party for kids doesn't seem quite right. The kids would probably be upset and the party a disaster.

The same steak has more value at the fancy restaurant than at a kid's birthday party. The steak doesn't change, but its value does. Why? A steak at a fancy restaurant helps you have a better restaurant experience. It delivers progress. A steak at a child's birthday party does not make the party better. It does not deliver progress.

This is why we say progress defines value, and contrast reveals it. You understand the value customers place on a product when you compare and contrast the progress it delivers against the progress other products can deliver. A steak makes a fancy dinner better but a kid's birthday party worse. A pizza makes a fancy dinner worse but a kid's birthday party better.

Solutions for Jobs deliver value beyond the moment of use. Imagine you own a car. When it's sitting in your garage, is it still delivering value? Doesn't the satisfaction of owning a car extend beyond when you're actively using it? What's more valuable: getting transported from point A to point B or having the peace of mind that you have the ability to go where you want to go, whenever you like?

Our lives are dynamic. They can't be measured well in static terms. Yes, a solution can provide functionality only in the moment, but its value to the customer is realized in contexts beyond that moment. A product should be designed with an understanding of how it improves customers' lives, not just how it offers value in the moment.

Producers, consumers, solutions, and Jobs should be thought of as parts of a system that work together to evolve markets. What is a system? A system is a collection of parts that work together to achieve a desired effect. The value is not in any one particular part of the system but in how those parts work together.

A car is an example of a system. Imagine I give you a box that contains all the parts of a car. What I gave you would likely be worthless to you. The parts are valuable to you only when they are assembled in a particular fashion, when they work together in a particular way, and when you can use them

What Are the Principles of Customer Jobs?

32

to make progress. You become better at helping customers, not by studying the individual parts of the car, but by studying how those parts work together to create something that helps customers make progress.

The same is true for producers, consumers, solutions, and Consumer Jobs. You need to understand how these parts work together to evolve customers and renew markets. Such a study will also help you understand why and how customers and markets don't evolve.

Grill manufacturer Weber understands the idea of producers, consumers, products, and Jobs as part of a system. Weber understands that it's not in the business of making and selling grills. It's in the business of making people better grillers. That's why it offers educational materials, recipes, party-planning guides, grilling accessories, and even a free phone hotline for grilling advice. For many grillers, the JTBD is also about entertaining friends and family with cooking theater, as well as tasty food. In this case, it's about becoming a better host and entertainer. Weber understands that no matter how well its grills function, if customers can't use them to make progress against their JTBD, the grills are worthless.

The understanding that customers are buying a better version of themselves is why Weber delivers a constellation of products that work together—as parts of a system—to evolve consumers and markets. Weber has been a successful, profitable company since 1893.

33

PART II

DEMAND AND COMPETITION

Customer Jobs theory encourages you to understand how demand for a product is generated and how customers view competition. The first three chapters in this part feature case studies of innovators who developed this understanding, and how it helped them create and sell products. Use these to create a mental catalog of examples of what it is like to apply Customer Jobs to innovation efforts. This will help you absorb the concepts in this book and become better at applying the theory to your own innovation efforts.

After these case studies, we'll dig deeper into the forces that shape customer demand, why Customer Jobs practitioners claim that Jobs remain while solutions come and go, and what it is like when an innovation effort fails to account for the forces of progress and how customers see competition

34

4 Case Study: Dan and Clarity

What's the JTBD?

Put it to work

I didn't know who Dan Martell was when I started writing this book. Another Customer Jobs practitioner told me about Dan's success as a serial entrepreneur, angel investor, and Customer Jobs practitioner. When I did catch up with him, I learned that he had applied Customer Jobs principles while building a company called Clarity. Customer Jobs thinking helped him:

Improve his research efforts
Understand the company's profit potential
Understand how Clarity could stand out to customers
Find marketing messages that resonate with customers
Know which features his team should—and shouldn't—add
to his product so that more customers would use it.

Founded in 2012, Clarity is a marketplace that connects entrepreneurs with experts who can advise, motivate, and inspire. Dan created Clarity to ensure that entrepreneurs get the advice they need to grow their businesses. It helps them find the right experts and then schedules and hosts calls with them. (Three years later, Dan sold it to Fundable, which is a platform entrepreneurs can use to raise money.)

Dan first heard about Customer Jobs from Eoghan McCabe during Clarity's early years. Eoghan is CEO of Intercom, one of the companies Dan invests in. Dan, intrigued by Eoghan's recommendation, believed Customer Jobs could help him grow Clarity faster:

Once I decided I wanted to learn more about Customer Jobs, the first thing I did was to search Clarity's marketplace. I found some [Customer Jobs] experts and did a few calls. It

Case Study: Dan and Clarity

35

was really helpful to get real-world experience and advice on how to approach it.

How can Customer Jobs help you do better research? Dan had already been a strong proponent of customer interviews, even before getting into Customer Jobs. Every week, he would call six customers or so and ask questions such as, "How would you feel if you could no longer use this?" or, "How can we improve Clarity?" But Dan knows that such interviews have limitations. In particular, he understands the difficulties inherent in talking with customers about their habits and that people often want to feel as if they are giving the "correct" answers. "I feel like customers have this really bad habit of lying sometimes," he said. "They'll say, 'Yeah. I love your product. I use it all the time.' Then, you look at the logs, and you realize they haven't logged in once since signing up—so you know it's not true." Calls with JTBD practitioners helped Dan realize the benefits of framing an interview around what Jobs customers are trying to get done. He did this by changing his questions. Instead of "How would you feel if you could no longer use this?" he asked customers, "Can you tell me about the other solutions you've tried? What did or didn't you like about each one?" In other words, he shifted from asking broad, individual questions to asking questions aimed at understanding customers' journeys as they searched to find solutions that fit their JTBD. He would then investigate if other customers had similar journeys. Dan said,

What I love about JTBD is that it really helped me to build a framework for those interviews. Before I became familiar with JTBD, I studied interview questions, extracted pain points, customer language, and all these other things. But when you frame it around the question, 'What is the Job your customer is hiring you to do?,' then it really puts a lot of things into perspective and helps you uncover key insights.

What do consumers consider as competition? How do you understand what customers do and don't value in a solution? Dan's new approach to interviewing customers encouraged him to learn about other ways they had

Case Study: Dan and Clarity

36

tried to get advice. He also wanted to learn if getting expert advice was really what customers were looking for. "Getting expert advice" is just an activity—a solution for a Job. What was the Job itself? What was the emotional motivation to make the customers' lives better? Answering these questions would help him continue to improve and promote Clarity.

To help guide him through these interviews, Dan kept asking himself a few simple but powerful questions:

What do customers see as competition to Clarity?
What would they spend their money on if they didn't spend it on Clarity?
Have customers set aside a budget for using Clarity or some other solution?

He then asked customers questions such as the following:
What other solutions did you try before deciding on Clarity?
What did and didn't you like about other solutions you had tried?
If you could no longer use Clarity, what would you use instead?

These questions helped Dan learn what his customers considered as competition to Clarity. He learned that before ending up with Clarity, customers had tried solutions such as joining entrepreneur groups, hiring individual advisers (who take equity), using LinkedIn, and attending conferences. "Understanding how people thought about our product and its competition helped us position it to be different," Dan said. "A lot of people had tried LinkedIn before coming to Clarity. Whereas LinkedIn connects people, it doesn't let them call in real time. It was also interesting to hear that customers considered Clarity as an alternative to attending a conference."

How do you learn what pushes customers to make a change? Dan began to learn two important observations as he talked with customers about the

Case Study: Dan and Clarity

37

solutions they had used: what his customers did and didn't value in a solution, and what was pushing them to make a change. He found these data by comparing and contrasting all the solutions they had used and asking himself, "What do these solutions have (or what do they not have) in common?" Dan realized that the solutions "use LinkedIn," "hire an adviser," and "attend a conference" had an important aspect in common: entrepreneurs were trying to make a connection with a specific person.

Dan and his team saw that entrepreneurs seeking advice valued the messenger, often more than the message. When it comes to getting advice, it's not just about the content. It has a lot to do with who's delivering it. Dan said,

There's real value in going after that person who is going to motivate you to make a change. It's not just having someone tell you, 'Go get ten sales tomorrow.' It is having billionaire entrepreneur Mark Cuban tell you, 'Go get ten

sales tomorrow.'
Dan now knew what was pushing customers to seek a solution:
entrepreneurs who were in a slump wanted inspiration from a particular
person. Getting advice is just an activity. If the seekers merely wanted advice,
they could have read a book or watched a video. They wanted more. They
were hoping that someone else's success would rub off on them. This is why
they wanted someone they respected to inspire and motivate them to get out
of an entrepreneurial slump. That was their emotional motivation to make a
change. Making progress with this Job is more valuable to these customers
than getting advice. Dan said,
I've got a list of competitors that tried to build competing
solutions. Their marketing and positioning was all about,
'Oh, if you want to talk to this type of person, we have
them.' But it was never about a person having the
knowledge. It was what [you knew] the person you talked
with had accomplished.
How can understanding the customer's moment of struggle help you market
a solution? These insights helped Dan and his team make two changes to
Case Study: Dan and Clarity
38
how they advertised Clarity. Each change would differentiate their solution
from what customers considered its competition and help customers realize
that Clarity was better. The first change was to emphasize that Clarity would
serve its customers on demand. As Dan put it:
We started saying Clarity gave "on-demand business
advice." It was adding the words on demand that
differentiated us from LinkedIn—which is an e-mail
exchange from which you may or may not get a response. It
also differentiated us from attending a conference—you
didn't have to wait until the next one came up. We
mentioned all that in the copywriting.
The second change was to highlight the fact that using Clarity was cheaper
than attending a conference. Dan said, "Understanding what customers
considered as competition also helped us position Clarity against the cost of
going to a conference. Why invest thousands of dollars in expenses and the
cost of a ticket if you can just talk to the speaker today?"
How did the product attract more users and customers? Clarity is a
marketplace for connecting buyers (entrepreneurs looking for advice) with
sellers (those offering advice). This means that Clarity needed to attract two
different groups of people—each with its own motivation for using Clarity.
The motivation for advisers is simple: they want to make money by
helping people. The entrepreneurs who use Clarity, however, are different.
They want to be motivated and inspired, usually by a particular person. This
meant that for its marketplace function to work, Clarity had to find experts
whom customers recognized and respected. Dan said,
Understanding what customers were trying to achieve with
LinkedIn and conferences helped us with the supply side of
the marketplace. We said to ourselves, "OK, if we recruit
experts, we need to recruit a certain type of expert." One of
the creative solutions that we came up with was to source
experts from SlideShare (a website where conference
speakers share their presentations with the public). If you
think about it, people who are regarded as inspiring and
Case Study: Dan and Clarity
39
motivational are those who give creative presentations at
conferences. When we wanted to add topics or categories
for Clarity, we would source experts who had presentations
on SlideShare.
How did Clarity realize its revenue potential? It was Dan's understanding of
what customers considered as Clarity's competition that also helped him
realize Clarity's revenue potential:
We learned from customers that their budget for Clarity

wasn't coming from the IPO, or from a monthly
membership, or from a training budget. It was coming from
spending money to go to an event to meet people and to
learn.

Dan realized that Clarity wasn't taking money away only from lower-cost
alternatives, such as LinkedIn, or from the price of a conference ticket. He
learned that Clarity was tapping into the budgets for big-ticket items, such
as hiring advisers and consultants, as well as entire budgets for attending
conferences, which include airfare, hotels, and meals. This explained why
his customers were willing to spend thousands of dollars on calls. This insight
helped him understand how valuable his product was to customers. It also
helped him understand Clarity's true value in case it was acquired, which it
eventually was.

Clarity discovered a silent competitor: anxiety. Nobody comes to
Clarity when he or she is having a great day. Dan and his team learned that
entrepreneurs were hoping they could get inspired by someone they
respected. Without this inspiration, these entrepreneurs would struggle to
put into action any advice they were given. This generated demand for the
product. But were there any forces that blocked this demand? Dan said, "The
biggest competition for us is when a customer chooses to do nothing. I think
that's true for a lot of innovations. In Clarity's case, entrepreneurs and
innovators continue struggling in the dark. They wouldn't choose to become
a self-educator and solve their problem."

Case Study: Dan and Clarity
40

Dan began to learn about the anxiety that blocks people from using the
product or keeps them from using it more, even when they do decide to
reach out to an expert on Clarity:

One of the questions that I would ask, which was about
Clarity as a solution and not their JTBD, was, "What can
we change to better meet your needs?" We found a bunch
of anxieties around using Clarity. A majority of them were,
"What if the expert doesn't answer my question? What's
your guarantee? Is the call going to be recorded? What
should I do to prep?" That last was one that really threw me
off. Both the seekers and experts themselves felt we should
teach them how to prepare for a good call.

Dan and his team had taken it for granted that people would be prepared for
a call. He assumed that both parties would set up the topic and then have a
conversation. This was partly true. Customers had specific questions, but
they didn't know how to organize them or what made a question good or
bad. Both sides wanted to prep, but they didn't know how.

Another anxiety that both parties shared was the fear of sounding stupid
or not putting their best foot forward. What if an expert doesn't have a good
answer for a question? What if he or she temporarily forgets the best answer?
What if I get nervous and forget my follow-up questions? These anxieties
prevented both groups—entrepreneurs and experts—from using Clarity
more.

How can you reduce the anxiety customers face when using or buying
your product? To fix the problem, Dan and his team added some prep
questions and guidelines to the e-mail templates they sent out to notify both
parties of a call. They also provided notes that outlined what a great call looks
like and what expectations the parties should have going into the call. Dan
said,

Discovering anxieties like those—that is the interesting part.
What I love is thinking, 'Here is the problem, and here is
the anxiety point. How do we solve it in a way that's
elegant, simple, and doesn't confuse the interface?' That was
always the fun part for me.

Case Study: Dan and Clarity
41

How can JTBD be used to research new features? As Dan became more
familiar with JTBD, he began to develop his own tools and processes that

would help him apply JTBD principles to improve Clarity. One such process was aimed at helping his team quickly validate ideas for new features. Before committing to developing a feature, the Clarity team wanted to make sure the problem they intended to solve was actually one that customers had. The best way to learn this was to find out if customers had tried to solve the problem before. Dan said, "An interview about how customers had tried to solve a problem in the past was more like a featureusage time line than a purchase decision."

An example of a feature the Clarity team chose not to build was one that saved search results when users looked for experts on Clarity's marketplace.

We asked customers questions like, "Have you ever tried to save results when you searched for an expert?" If they said no, then we'd move on. We then asked, "Do you have a browser bookmarklet? Which ones?" Then, they would say, "Evernote, Buffer…" It would provide so much context outside of the feature. It was more about how the customer had tried to solve their problems in the past.

So, Dan's team decided not to build the browser bookmarklet. They didn't think it delivered enough value because the problem it solved wasn't one their customers had struggled with. Dan said,

A lot of people—especially if they're committed or already invested in a solution—are looking for that confirmation bias that it's something they should do. It's a different question to ask customers how they solved the problem in the past. I could ask them, "Hey, what do you think of this?" They might say, "Oh, it's prettier. It works great." But that's not really answering the question we're asking. We want to know, "Are you going to use it? Are you struggling to make progress? Have you tried to solve this in the past? Do you want to hire someone or something to

Case Study: Dan and Clarity
42

solve this Job to Be Done?" If the answer is no, then cool. We write that down and move on.

How does JTBD help innovators? Dan appreciates the focus JTBD puts on exploring customer motivation. He also wishes more companies would do that rather than "spy" on customers.

I think the biggest thing that Jobs [JTBD] encourages people to do, which I'm a big fan of, is to stop spying on customers and start talking with customers. I feel that way especially with software because we have the analytics and the geeks who are building the software; they're all about tracking and logging and all these data…I always give the analogy of being a retail shop owner and hiding in the back room and trying to learn from your customers by watching the closed-circuit television.

You could watch [customers] come in, walk around your store, pick up things, put them down, try things on…or you could just walk out and ask them, "Hey, what brought you in here today? What are you looking for? What other places did you try in the past?" Talking to customers about their motivations is where you're going to learn the most.

WHAT'S THE JTBD?

From the data Dan has given us, I'd say that the desire for progress is as follows:

More about: getting out of a rut, making a connection with someone whose accomplishments I respect, being inspired, being motivated to act, feeling like I'm on the right path, having confidence in what I'm doing, having success rub off on me, on demand

Case Study: Dan and Clarity
43

Less about: getting expert advice, talking with an expert,
giving away equity, having a video chat with a mentor, emailing
a mentor, mentoring, meeting other entrepreneurs,
seeing a mentor live

Here are some possible descriptions of the one or more Jobs to be Done
Clarity is hired for:

Help me get out of an innovation slump with inspirational
advice from someone whom I respect.

Give me the motivation to act with a kick in the butt from
someone I respect.

Take away the anxiety of making a big decision with
assurance from someone else whose has been in a similar
position.

These work for me because they don't describe an activity or task. They
describe the motivation that comes before engaging in an activity (i.e., using
a solution). Also, notice how these descriptions can be used to describe the
other solutions customers had tried in the past (e.g., attending a conference,
giving away advisory shares, and using LinkedIn). This is important because
a JTBD either doesn't change, or does so slowly. If a description of a JTBD
works for solutions from one hundred years ago, it'll probably work for
solutions one hundred years into the future.

PUT IT TO WORK

Dan's case study is a great introduction to Customer Jobs. Here are some
suggestions to help you get started today with applying Customer Jobs
thinking.

Ask customers about what they've done, not just what they want.
Confirm it if you can. Customers will often tell us what we want to hear,
even if it's partially (or completely) untrue. Customers may tell you that they
use your product "all the time," but they really use it only intermittently.

Case Study: Dan and Clarity
44

Also, people build easy-to-remember narratives between themselves and the
products they use. Phenomena like this are why it's tricky to ask customers,
"what do you want?" and "How can we make things better?"22

The answer for these problems is to talk with customers about what they
actually did, not just about what they say they want. What were their
revealed preferences, not just their stated preferences? Even the answers
about actual action taken won't be 100 percent accurate, but they will be a
great deal more reliable than their answers to what-if questions.
Understanding how customers have solved problems is a crucial part of
understanding their JTBD. Not only does it help you understand what
customers expect from a product, but it also helps you design features for
new products.

Ask the right questions to learn how your customers view competition.
Accurate models of competition can come from only customers. Any model
of competition that doesn't come from them is invalid. One way of getting
the information you need to build a correct model of competition is through
customer interviews and surveys. Ask them questions such as the following:

What other solutions did you consider before trying the
product?

What other solutions have you actually used?

If the product wasn't available to you, what would you have
done instead?

What solutions have the people you know tried or used?

Learn what kind of progress customers are seeking. What's their emotional
motivation (JTBD)? Use that to segment competition. Dan learned that
Clarity's customers saw its competition as attending conferences, using
LinkedIn, and hiring advisers. These solutions have vastly different
functionality and qualities. However, from the customers' point of view,
they appeal to the same aspiration: "Get me out of an entrepreneurial slump
with motivational advice from someone whom I respect." Discover your
customers' motivation through comparing and contrasting the solutions that
they consider as competition:

Case Study: Dan and Clarity

45

What do the various solutions have in common? What is
different about them?
What did or didn't the customers like about each solution?
What would customers do if they couldn't use their existing
solution for their JTBD?
What would the consequences be?
How are they expecting life to be better once they have the
right solution for a JTBD?
These types of questions help you understand two things: what customers
are struggling with now, and how they hope life will be better when they
have the right solution. Put these two together, and you'll have their JTBD.
Ask yourself, "From which budget will my product take away money?"
Also ask, "When customers start using my product, what will they stop
using?" Dan learned that his customers were willing to spend thousands of
dollars on Clarity calls. This number didn't come from looking at how much
other "talk to an expert" products cost. He learned this by understanding
that his product—from the customers' point of view—was replacing the
entire budget of going to a conference.
I've noted that when it comes to solutions for a JTBD, customers can
use only one at a time. When they start using one solution, they have to stop
using something else. This helps you understand what the competition is. It
also helps you gauge how to price your product properly and figure your
revenue potential. Should you charge less or more? You have two big factors
to consider: the amount customers are already accustomed to spending on a
solution for a JTBD, and the intensity of their desire to change. The more
they hope to change, the more they are willing to pay.
Create better marketing material by speaking to your customers' JTBD.
Dan Martell described Clarity as "on-demand business advice." He also
featured access to experts whom customers would recognize. He also
positioned Clarity as an alternative to going to a conference: Why spend the
time and money going to a conference? Talk with the speaker today! Both
of these messages spoke to customers' motivations and distinguished Clarity
as unique.

Case Study: Dan and Clarity

46

Talk with customers to learn what messages connect with them. It can be as
simple as asking them to describe why they like your product. Sometimes,
you have to probe deeper and ask them questions such as, "Before you
bought our product, how did you know it was right for you?" The best
promotional material, however, comes from speaking directly to their desire
for progress.
Focus on delivering emotional progress (getting a Job Done). Don't
focus solely on functionality. Dan mentioned a list of people who had tried
to create solutions similar to Clarity. They failed, and Clarity won because
Dan designed and marketed it in a way that spoke to customers' emotional
motivation. The Clarity clones thought of themselves as "talk to an expert"
products; they were focusing on functionality, activities, and tasks. But Dan
focused on the emotional quality—that is, customers' JTBD. He knew that
customers wanted to be motivated and inspired by someone whom they
respect. This made Clarity stand out, and it's why Fundable acquired it.
Clarity's former competitors, however, have already been forgotten.
Your guiding star in understanding your customers' JTBD is their
motivation to better their lives. Focus on that. Focusing on functionality will
distract you.

47

5 Case Study: Anthony and Form Theatricals

What's the JTBD?
Put it to work
What progress might someone use theater for? I had never asked this
question before, but Anthony Francavilla had. For the past few years,

Anthony has been applying Customer Jobs principles to figure out the answers to that question. Theater has been around for thousands of years. Shouldn't we know the reasons why people attend the theater? Maybe. But maybe not.

Anthony has managed and produced theater for ten years. In 2012, he cofounded Form Theatricals, whose mission is to help productions grow and be successful. This is particularly challenging in the theater world. Many productions are run by actors or writers who often don't have much business experience. They also have little to no experience innovating around customer motivation. This is where Anthony and Form Theatricals come in. Customer Jobs has helped Anthony figure out how to learn what really matters to theatergoers; what customers do and don't consider as competition to theater; and how a theater production could improve its shows for patrons, increase profits from ticket sales, develop new types of theater products, and reduce the cost of operating a show.

Why look into Customer Jobs? Anthony knew that interviewing theater patrons was the key to improving a show. But what's the best way to interview people about a show they've just seen? To find out, he sought advice from someone who specializes in interviewing customers. Anthony said,

I got together with this guy, Boris, who specializes in ethnographic interviewing. I said to him, "I have this problem with a client. People don't like the show, but it's selling well. I want to interview customers, but I don't know what I should be asking them about." He said I could talk to them and try to find out what Job these patrons are trying to get Done. He asked me if I had heard about Customer Jobs. I told him I hadn't. He explained it to me. Then, he told me about some sources online where I could

Case Study: Anthony and Form Theatricals
48

learn more. I also signed up for the Customer Jobs Meetup that's run here in New York.

After looking into Customer Jobs a bit more and learning about some of the tools associated with the principles, it didn't take long for Anthony to start gathering powerful insights.

Studying what customers consider as competition helps you reveal what pushes them to change. It also helps reveal their JTBD. Anthony applied some Customer Jobs thinking to his next client: a children's theater company. To begin, he interviewed parents who had taken their children to the company's show. He wanted to know why they chose this particular show. Did they consider any other activities for their children besides attending the theater? He told me, "We interviewed a bunch of parents. We learned that the options they had considered [as alternatives to attending the theater] ranged from going to The LEGO Movie and buying the LEGO video game to signing their children up to clubs—like the Girl Scouts."

This was a story Anthony kept hearing. Parents were considering a wide array of options as alternatives to taking their children to this particular theater show. Or, in Customer Jobs terms, he learned exactly what parents considered as competition for their JTBD. These customers surely used the theater for other Jobs in other circumstances. But in this case, what Job were they hoping to get Done by bringing their children to this show?

To help him answer this question, Anthony applied Customer Jobs's idea of "contrast reveals value." He talked with these parents about what they did or didn't like about the other options they had considered. What can the theater do for them that an alternative solution—such as the Girl Scouts—can't? He also asked these parents about what they did immediately after the shows. Did they have family discussions about them? What were those discussions like? After talking with numerous parents, he began to see a distinct pattern. "We found out that part of the Job these parents are trying to get done—when it comes to entertainment and activities for their kids—was that they wanted help teaching their kids how to be independent…while also reinforcing that they are a member of a team."

How do you go about making product changes when you understand
the customers' desire for progress? Anthony brought these insights to his
client. Together, they decided to rewrite parts of the play. They kept most
of the content the same, but they added a story arc wherein the hero works
Case Study: Anthony and Form Theatricals
49
with the characters around him to solve a problem. This would give parents
a talking point with their children about the importance of working with
others. Anthony said,
It's interesting to me because helping writers understand
what Job parents are using their play for is more powerful
than saying to them, "Write a movie, or write a play that a
nine-year-old will like." When we know that parents have
a Job that involves their desire to teach their children life
lessons in an entertaining way, we can work with our clients
to help them craft their content better. When we present it
as a Job to Be Done, the artist has a lot of leeway around
what the story should do.
For Anthony's client to sell more tickets to these parents, the performances
had to help parents make progress against their JTBD, which included their
aspiration to be responsible parents, as well as becoming better at teaching
life lessons to their children. This needed to be done in a way that their kids
would enjoy. The performance also had to do this better than what parents
considered as competition—namely, other plays, movies, video games, and
clubs.
Anthony wouldn't have got the same depth of insight had he
interviewed parents only about what they did and didn't like about the play.
Had he done that, he would have ended up getting a lot of feedback about
how to make the play better—but only in comparison with other plays. With
a Customer Jobs approach to understanding competition, he was now
learning how theater compared with other solutions customers had tried.
What do we gain from digging deeper into the JTBD? Anthony wasn't
satisfied with just the one insight that these parents wanted help teaching
their children life lessons. He wanted to dig deeper into their motivation.
Were there other ways they were hoping that theater would make their lives
better? He said,
One of the things we figured out was that parents want to
have shared experiences with their kids. That's not
necessarily understood by the producers of theater and
Case Study: Anthony and Form Theatricals
50
movies. On the surface, it doesn't seem like a shared
experience. Theater productions often see the dynamic of a
play as "Let's just go sit in a dark room and watch this
together." What we learned—and what a lot people don't
realize—is that the shared experience actually happens after
the show. It's when everyone goes out for dinner and they
talk about the movie or the play they just saw.
This insight about shared experiences prompted Anthony to ask parents
other questions. What were other shared experiences they engaged in with
their children? How did theater fit into those?
I interviewed this father about how he, his wife, and his
child would pick what they were going to watch on TV.
They were basically engaged in rhetoric; they would each
debate what they'd wanted to watch. They'd go back and
forth, to the point that sometimes the debate would end
with them all deciding to just go their separate ways and
reading their own books. They wanted to have a shared
experience—to the point that the debate itself became the
shared experience—and they didn't end up watching
anything on TV.
By comparing and contrasting how families had and felt about shared
experiences, Anthony could begin to understand what customers did and

didn't like about each solution. What made discussion about what to watch on TV so successful? What things didn't families like about it? What were family discussions like after the family saw a play together? Were these discussions about life lessons or about other things, such as the performances of the actors? How could a theater show promote better conversations at home?

Answers to these questions helped Anthony understand that these parents wanted to make family life better through engaging and educational discussions with their children. These conversations were a bonding experience. This was exactly the kind of direction his client needed. It helped

Case Study: Anthony and Form Theatricals

51

the children's theater company make script adjustments so its plays could act as vehicles for family conversations.

How many Jobs might an innovation be used for? Anthony's interviews with families had been successful. Understanding what Jobs they were using theater for helped him provide guidance for his client. He decided to continue Customer Jobs research with his other clients.

The next few shows he worked on were drama pieces with more serious subject matter—definitely not for kids. Patrons were usually individuals or small groups of friends. What Job might these people be using theater for? Anthony said,

We interviewed a banker who went to a show by himself. He said, "I love these weird off-off-Broadway plays." As we dived deeper into what that meant, we began to realize that an important part of the theater experience was who else is in the audience. That's what one group of customers was looking for. They would say, "I want to hang out with artists more." Others would say, "It's just amazing. I don't normally sit in a room and have an experience with a group of diverse people like that."

This is how Anthony began to discover another Job that people use theater for: it was about being a part of, or dipping their toes into, a different community. Very often, these customers had careers that weren't arts related, such as banking or law. He would hear comments such as "I like these productions that are a bit out there," "I like going because there are artists in the audience. They're talking about art," or "I don't have a job in the arts, but I love the arts. I want to be involved in that kind of scene." For many of these patrons, going to these shows was their only opportunity to engage with a diverse group of people. They liked the arts. They wanted to be involved in that community. Anthony said,

That was a very impactful insight. A lot of times, the theater will try to sell you the idea that it's like a movie—but on stage. You can't compete with that. Theater is more expensive. It's sometimes super-inconvenient to attend a

Case Study: Anthony and Form Theatricals

52

show. I have Netflix. If I want to watch a movie, I can hit a button, and there is a movie.

Figuring out what Job live entertainment solves for people in the twenty-first century is exciting. We've learned that, yes, it is entertainment, but it's also about this idea of community. It's something that you're going to enjoy with other people. Maybe there will also be drinks, food, a lively atmosphere—all that kind of stuff. That's something that a theater can take and use to build up a new business model for the twenty-first century—as opposed to this idea that there's going to be a celebrity in the show. Tickets for those shows are two hundred and fifty dollars. There's a very limited audience for that.

How can Customer Jobs help you reimagine existing products? With these new insights, Anthony and his clients were able to create a new type of theater experience: a theater subscription product. When people buy these subscriptions, each person is put into a specific cohort of customers. Shows are picked out for these customers. Over several months, this same group of people sees the same shows and engages in social events around the show. Anthony said,

These patrons valued this idea of inclusiveness. It is important for us to help theater productions understand that patrons are looking for an inclusive experience. This meant making the subscription affordable. It's easier for a banker to pay five hundred dollars for a few shows than it is for many artists. We solved this problem by offering multiple payment options. You could spread the price over months or pay for the whole subscription up front.

How can Customer Jobs help you avoid wasting resources by building features that customers don't care about? Anthony didn't help his clients by only suggesting to them what they should add to a show. He also made suggestions on what to take out of it.

Case Study: Anthony and Form Theatricals

53

One of Anthony's clients had a production that featured an after-the-show tour of the stage for anyone who attended. It was something the producers of the show were proud of. But did patrons enjoy it? As he interviewed patrons after the show, he learned that most of them hadn't known about the tour when they bought their tickets. They had simply chosen the show because tickets were being sold at a discount. He said,

The majority of these patrons learned about the show and the discount on the day of the show. For some of them, it had been a last-minute decision. They'd be discussing with friends what to do for the evening. Should they just go to a bar? A comedy club? But when they noticed the discounted theater tickets, they then chose to buy tickets. It could be an hour or two before the show started.

Of all the people he interviewed, only one or two knew that the set tour was going to happen. Anthony's client had assumed that theater patrons were interested in access to the actors and seeing how the show worked. As it happened, almost no one who bought a ticket knew about the tour. The tour hadn't been part of these patrons' purchase criteria, so it didn't help explain why they were hiring the show. Anthony said,

we learned that people were not hiring the show to get access to the actors and set after the show. Finding that piece of information was very valuable. The after-show tour was expensive to maintain, and it wasn't something patrons were particularly interested in. The Job for those patrons was about entertainment and having a shared experience with their friends and significant others.

In this case, the producers had overengineered the show. They had designed the show based on what they valued—a tour of the set—instead of what their customers valued—having a shared, fun experience with their friends. After gaining these insights, Anthony worked with the producers to discontinue the set tours. While experimentation is good, it has to be within the constraints of the Job that customers are hiring the show for. The new

Case Study: Anthony and Form Theatricals

54

thinking freed up the show's designers to focus on what they were doing right and make that better.

How does anxiety stop customers from buying your product? Is there really such a thing as an "impulse purchase"? Similar to tickets for airplanes, sporting events, and movies, theater tickets are worthless after the event starts. Seats are perishable inventory. This posed an interesting challenge for Anthony and his clients. To help him figure out how his clients could sell more tickets, he began interviewing customers to learn more about the key events that sped up or slowed down a decision to buy a ticket. Were there any anxieties about attending a particular show? If so, how could a theater production solve this problem? Anthony said,

For each customer, we mapped out a time line of the events
that led up to their ticket purchase. We began to hear the
same things over and over again. Things like, a husband
reads a magazine with his friends at work—that's where
he'll first find out about a play. He'll e-mail his wife about
it. She'll respond with a comment like, "Seems interesting. I
like that it's a horror-themed play set in Spanish. I like
horror." But when bad reviews for it come out, they start
to doubt if they'll like it. But they still keep an eye on the
show. Then, maybe a week later, they'll learn about the
discount. At that minute, they're pushed over the edge and
buy the tickets.

Anthony discovered two insights here: some anxieties prevent customers
from buying tickets to a show, and tickets that seem to be impulse purchases
sometimes aren't.

The majority of customers who bought a ticket through a discount did
so on the day of the show, but that doesn't mean these were impulse
purchases. In the backs of their minds, these customers already had specific
shows they wanted to see. But what was holding them back from buying the
tickets? Anxiety. They'd first be excited about a show's concept, but if
reviews weren't positive, they'd hold off. The discount, however, could
compress the purchase time line. It eased anxiety and caused potential patrons
to buy.

Case Study: Anthony and Form Theatricals

55

Can Customer Jobs theory gather new insights about a medium that is
thousands of years old? As the competition for theater changes with the
advance of technology, it's important to focus on the Jobs that customers
hire theater for. Many parents use it as a way to help them have the types of
conversations they want with their children and to help them teach life
lessons. For those who want to expand and bring diversity to their social
circles, community and diversity are critical.

Anthony's application of JTBD principles and focusing on customer
motivation have enabled him to innovate within a medium that is thousands
of years old.

WHAT'S THE JTBD?

This case study reveals different directions of progress that people hope to
make using theater. This would explain why there are so many different
types of theater shows. Some big themes associated with Jobs to Be Done I
heard include using shared experiences to create or strengthen bonds with
family and friends, parents teaching their children life lessons, and adding
excitement to your social life by getting involved with people whom you
normally wouldn't interact with.

The clearest JTBD I heard was related to parents' desires. They wanted
to teach their children how to be independent, while also understanding
how to work with others. This works for solutions such as video games,
movies, clubs like the Girl Scouts, and attending the theater. This case study
had some great data about customer motivation; however, I still have
questions about these parents' motivations:

What are some of the consequences of not teaching their
children life lessons?

Is there something in these parents' lives that is pushing
them to make a change now, or are they deciding to be
proactive and avoid feeling guilty in the future?

Does having conversations about life lessons relate to
anything else going on in the lives of these parents or
children? What about school or interactions with their
friends?

Case Study: Anthony and Form Theatricals

56

What other solutions do parents couple with theater to
make progress?

How will parents know their Job is Done? That is, when do

they know they are making progress and things are getting better?

I would have a better idea of what progress parents are trying to make once I had answers to questions such as these.

PUT IT TO WORK

How do you persuade teammates or management to change a product?

Frame design challenges as a JTBD. Innovators like to solve problems; we don't like being told what to do. I find it's best to motivate a team by presenting them with problems to solve in the form of a customer's JTBD.

Dig deeper when you tap into a struggle or aspiration. How have customers tried to solve it before? Anthony discovered that parents had a desire for shared experiences with their kids. But what does shared experiences mean? It turns out that a shared experience is most important after the show. This insight gave Anthony the idea to talk with other patrons about their shared experiences. What made a shared experience successful? How had the patrons tried to have shared experiences?

When customers describe a struggle or aspiration, don't make assumptions about what they mean; rather, unpack what they're saying. Ask for specific examples. If they describe a struggle, how do they imagine life being better once they solve it? If they describe an aspiration, what are the consequences if they can't achieve it? The answers will help you make design, marketing, and business decisions.

Discover what customers value. Learn their expectations at the moment of purchase and/or first use, and avoid overengineering solutions. Anthony had a client who offered a costly after-the-show tour of the set. However, he learned that almost no patrons were aware that the tour was being offered, so it didn't affect their purchase decisions. This made it safe to remove tours from the show. This reduced costs of production, and it increased profits.

Case Study: Anthony and Form Theatricals
57

A great deal of waste happens when solutions are developed with features that customers don't value. Customers value the progress a feature may deliver, not the feature itself.

If you have an existing product, engage in an audit to determine which features don't help customers make progress toward their JTBD. If you're about to create a new feature, make sure it delivers progress and, more importantly, helps you increase profits. You might learn just as one of Anthony's clients did—namely, that you're spending money to support features that customers don't find valuable.

Determine if anxiety is a competitor. If it is, find ways of reducing it. You should attack the anxieties in choosing and using a product with the same fervor as attacking a competing product. If customers have anxiety over the cost-value relationship of your product, offer a discount. If customers experience anxiety in using your product, find a way to make your product less intimidating. Anthony attacked the former by offering discounts on the day of the show. He attacked the latter by offering drinks as "liquid courage" for theater patrons to feel more comfortable mingling with one another.

Be suspicious of the "impulse purchase" concept. No purchase is random. Anthony discovered that many customers purchased tickets on the day of—or even an hour or two before—the shows. But that doesn't mean that these were impulse purchases. Many patrons had already decided they wanted to see a show; they had reservations about paying full price for a show that had received mixed reviews. A lowered price helped ease their anxiety about paying for a show that might not be very good.

Talk with customers about how they came to choose your product for their JTBD. They might claim that their purchase of a USB charging cable was "just an impulse purchase while I was waiting in line." However, when you dig deeper, you might learn that they were about to go on a trip and wanted to take an inexpensive charging cable with them in case it got lost during their travels.

58

6 Case Study: Morgan and YourGrocer

What's the JTBD?

Put it to work

Morgan Ranieri was fed up. Getting home from work at seven o'clock at night meant he couldn't get the groceries he wanted, for the stores he wanted to shop at were closed by then. Instead, he had to settle for the supermarket chains around Melbourne, Australia, where Morgan lives. I say "settle" because the food quality at these supermarkets isn't very good. Shopping there also meant he wasn't supporting family businesses, which was something he liked to do.

Sensing that he shared this struggle with other people, he teamed up with his colleague Bandith and created YourGrocer. The concept for YourGrocer was simple: have your groceries delivered to you from local, high-quality food shops.

Over the next few months, Morgan and Bandith did some tests to see if the YourGrocer concept could work. They investigated what the competition might be, what logistics would need to be in place, and how many local shops were interested in partnering with them, and they even did some preliminary testing with a few customers to get feedback.

Their tests told them that an opportunity did exist. However, to grow their business, Morgan and Bandith needed someone with more technical expertise to join the team. Morgan met Francisco (Frankie) Trindade at a local Meetup. Morgan said, "Over the next month or so, we began speed dating, in a sense—getting to know one another before deciding to work together."

In this case study, we learn how Customer Jobs helped Morgan build a consensus among team members, what customers did and didn't value in a solution, find the right marketing messages, how it helped first-time customers switch to Morgan's product, and how he could reduce churn. Customer Jobs helps you persuade others that an opportunity exists.

Frankie wanted to make sure an opportunity existed before he joined YourGrocer as its third cofounder. This is when Frankie introduced Morgan to Customer Jobs. Frankie told Morgan that he wanted to spend more time learning what Job(s) customers would use YourGrocer for. He especially wanted to do this before writing any of the software that would power the business. Morgan said,

Case Study: Morgan and YourGrocer
59

It was Frankie who introduced me and Bandith to Jobs (Customer Jobs). Actually, the first thing Frankie did when he joined YourGrocer was to make sure we all understood the principles of Customer Jobs. We spent a week learning about it and figuring out how we would interview customers. All of us learning about Customer Jobs, and then interviewing twenty customers together, was a great way to induct him as YourGrocer's third cofounder.

How does your team benefit from doing Customer Jobs research together? The newly formed YourGrocer team gained an unexpected benefit of doing Customer Jobs research together. As Morgan said,

This shared learning experience really helped bring us together. We developed a shared understanding of what the business needed to be—which was missing in the beginning. At the start, we all had very different ideas about what customers were struggling with and how we should solve it. I was the typical visionary cofounder who has the next five years planned out in my head—which is very dangerous. But Frankie didn't make many assumptions. He wanted to take it one step at a time. His middle name should be "Pragmatic": Frankie Pragmatic.

Interviewing about twenty customers got me, as a business cofounder, and Frankie, as a new technical cofounder, on to the same page.

Are data about "types" of people information or misinformation? The first aha moment for the YourGrocer team was when they realized that their customers didn't match the assumed demographic. Morgan said,

we had an assumption about what our customer
demographic was—or the idea of who our target customer
was. The reality turned out to be quite different. We
Case Study: Morgan and YourGrocer
60
thought we were creating a business for young professionals
who wanted to buy groceries online. It turns out, almost
every single one of our customers was a young family—
typically a young mom with a couple of kids at home.
At first, the YourGrocer team created the business out of their own need—
that is, a way for busy young professionals to buy groceries online. But
because most of the company's customers were young families, the team
needed to adjust. "It just turned out that the type of customers we were
targeting at first [young professionals] didn't really work too well for our
product, but this other group of customers [parents] was ripe for it."
How do struggling moments arise? what is it like to be pushed to
change? Morgan and his team had now picked up on a group of struggling
customers. The next step was to learn how and why these people were
struggling. what was the struggling moment? This meant that Morgan first
needed to talk with these customers about the different ways they had
purchased groceries before. Morgan began to uncover the triggering events
that would push these customers from one solution to another.
The push that eventually led our customers to YourGrocer
often began a couple of years in the past. They'd start off
shopping at the shops they liked. Then they'd have their
first child. Getting around to these shops with one kid was
difficult, but they could deal with it. But once they had
their second child, that would really change things. Having
a second kid made it almost impossible to get to the local
shops they wanted. That's when they switched from their
local shops to buying at the two big suppliers here.
As a customer's family grew, more of his or her time was dedicated to
caregiving. It also made traveling to multiple food shops difficult. This would
lead these families to consider other ways of getting their groceries, such as
at supermarkets.
Discovering these triggering events helped Morgan understand how
demand was being generated and how it pushed these parents to seek a
Case Study: Morgan and YourGrocer
61
solution. This helped him get an idea of how these parents were trying to
make their lives better—that is, what Job they were trying to get Done.
what is it like to learn what customers do and don't like about solutions
they've tried? Next, Morgan had to learn how these parents had already tried
to solve their problem—namely, how to get groceries when they had
children to take care of. Comparing and contrasting these solutions would
help him understand what these customers did and didn't value in a solution.
In particular, these parents complained about cost, poor-quality food, and
not being able to choose foods they wanted.
The big supermarkets do fresh produce badly. The other
local delivery suppliers that do fresh produce well are
expensive. Some of them even have these subscription
models where you get a preselected assortment of groceries.
Customers can't pick and choose what they want, when
they want it. Our customers didn't like that. They were
getting a bunch of stuff they didn't want, not using it,
having it all go bad, and getting frustrated by that. Often, all
these issues with other services had been going on for some
time. They were just putting up with it. Then, we came
along. It was just what they had been waiting for.
what are examples of things customers value? Before starting YourGrocer,
Morgan and his team already had a pretty good idea of what the business
would be: home delivery from the quality shops his customers loved. Now
they were filling in any blanks and confirming their assumptions of the value

that YourGrocer should deliver. Here's what they were learning:
Convenience had become these customers' top priority.
They used to value food quality the most, but traveling with
their kids to multiple stores proved too difficult for them.
This pushed them to trade food quality for convenience.
23
After convenience, they wanted to be able to choose foods
they wanted. This ruled out services that delivered to the
home but didn't allow buyers to choose their own food
options.
Case Study: Morgan and YourGrocer
62
Quality got pushed to the bottom. Ultimately, these
customers ended up choosing food from supermarkets.
while supermarkets offered the lowest-quality foods, they
ranked the highest on convenience and selection.
How does JTBD help you create a message that connects with customers?
The YourGrocer team members were confident that they now understood
what the customers valued and that the team could deliver this value. The
next step was to figure out a message that would connect with customers.
Once again, customer interviews helped Morgan and his team figure this
out.
In the beginning, we didn't know which messages would
stick with customers. We would say, "It's good to shop
locally, because it's good for the environment. It's better
food. It's better priced. It's convenient. It's local shops. It's
good for your community." We were throwing out half a
dozen different messages out there without knowing which
ones would persuade customers to try us.
Morgan solved this problem by asking his customers JTBD-style questions,
such as "what stood out to you about us?" As he did so, he began to gain
rich details about customers' motivations.
One thing we really like about Customer Jobs is that you
want to learn from customers what they've done in the past.
You're not just asking customers their opinion at the time
you're talking with them or through a survey. We would
ask them, "What did you tell your friends about
YourGrocer?" Or, better yet, "Can you show me the text
in your phone that you sent your friend about us?"
Morgan's customers had no problem pulling out their phones and showing
him the text messages they had sent to others about YourGrocer, as well as
any Facebook posts they had made about shopping with it. In those messages,
Case Study: Morgan and YourGrocer
63
Morgan saw customers express what they felt was valuable about
YourGrocer. He learned that the messages they sent to other people were
about getting groceries from a store:
We knew that customers wanted quality foods. But just
saying we offered quality wasn't enough. We learned that
customers trusted our message of quality only because of the
local stores we featured on our site and advertisements.
Other messages didn't stick with them—being good for the
environment, our competitive pricing, the ease of use when
compared to other delivery services. All that kind of stuff
wasn't really standing out to our customers. It turned out,
they were buying from us because they recognized the
stores that we featured on our website.
This is when the YourGrocer team honed their advertising message. It
combined convenience, variety, and quality into one statement: "Online
grocery shopping and same-day home delivery from the local shops you
love."
What anxieties do first-time customers experience? What might prevent
customers from using your product? So far, Morgan has learned about how
Page 35

important convenience is to his customers. Just how important this was
became even more salient when he talked with customers about the first
time they tried to use YourGrocer for delivery. He said,
We learned about this one anxiety: a lot of people came to
the site and had trouble trying to figure out how
YourGrocer would fit into their lives. We kept hearing
comments such as, "I just don't know when my groceries
are going to get delivered." This struck us as odd because
we give really flexible delivery options.
This anxiety didn't make sense to Morgan and his team. They offered flexible
delivery hours, so why were customers commenting about not knowing
when their groceries would be delivered? The answer lay in customers'
Case Study: Morgan and YourGrocer
64
shopping habits and expectations. Morgan said, "It turns out that customers
had this obstacle in their buying path. They decided what groceries to buy
only after they'd figure out when they'd get the delivery. We had it reversed:
you would pick your groceries first and then decide when to have them
delivered."
First-time customers coming to the site already had an idea
of how YourGrocer was going to work. They had a habit
or expectation of coming to a site, finding out how soon
they could get a delivery, and then deciding what to buy.
When this expectation was violated, they became frustrated
and anxious. At this point, they would abandon trying out
YourGrocer.24
To fix this problem, YourGrocer adjusted the checkout process. It asked
customers to pick a delivery window first and then walked them through the
grocery-selection process. "That helped," Morgan said. "We saw
conversions go up after that."
What habits prevent customers from making progress? Can customers'
habits be competition? Anxiety wasn't the only emotional force the
YourGrocer team members would face. They also had to navigate
customers' existing habits. Morgan said,
Dealing with customers' existing habits was definitely a challenge
with repeat-purchase customers. They had this habit around
being able to duck down to the local store when they ran out of
a key ingredient while cooking. Then, while they were at the
store, they'd pick up extra groceries. In this case, they wouldn't
need to come back to us for another two weeks. Sometimes
they'd fall out of the buying cycle, and we'd lose them as
customers. Habits like these are our biggest competition.
If Morgan wanted to keep customers coming back, he needed to make sure
that customers developed new habits around using YourGrocer. He couldn't
focus on only the outcomes customers were looking for. He had to think
holistically about the customers' JTBD. Customers didn't just want their
Case Study: Morgan and YourGrocer
65
groceries delivered; they wanted a solution they could use to make their lives
better.
So, how did Morgan and his team solve it? They focused on helping
customers become more successful at using their product.
We get people to set up regular orders with us. We set up
e-mail triggers to help remind them that they might need
something. The first one goes out three days after getting
your first delivery. We send you an e-mail saying, "Hey, do
you need a top-up on anything? Here's a free delivery of
any size so that you can top up with us." Seven days after
your last purchase, we e-mail you again and ask, "Do you
know how easy it is to repeat last week's order? You can
just click this button and get everything delivered again."
These e-mails are part of YourGrocer's efforts to help customers become
better meal planners. This is important to note because customers aren't

consciously joining YourGrocer to become better meal planners. It wasn't an outcome that customers were seeking; however, meal planning is what customers have to be able to do if they want to use YourGrocer for their JTBD.

What progress are customers trying to make? Morgan and his team came to understand their customers' JTBD by combining their own intuition with what they learned through customer interviews.

An important part of our customers' Job to be Done is, "Give me a way to provide quality food for my family without the stress of running around." The phrase "YourGrocer does the running around for me" came up quite a bit during the interviews. Before YourGrocer was available to them, if they wanted to go to these local shops, they had to be willing to deal with running around to these different stores—and deal with the hassle of having their kids in tow.

Case Study: Morgan and YourGrocer

66

Morgan had the first part of the JTBD: his customers were struggling to get quality groceries without all the stress. Next, it was time for him to understand how customers were expecting their lives to be better when they had the right solution. what would it be like when this Job was Done? YourGrocer helps families get back their Saturday mornings and weekends. With us, they can now buy good food for their family without having to sacrifice their Saturday mornings or weekends visiting all these different stores. That's the trade-off they were struggling with before. If they wanted quality food for their family, they'd have to give up some family time so they could go shopping. If they didn't want to give up family time, then they'd have to deal with poor-quality food from the supermarkets.

How can you beat the competition? Eliminate the need for the customer to make a trade-off. YourGrocer wins because it does what every great innovation does—that is, it helps customers break a constraint. Using YourGrocer means no longer choosing between quality food for the family and quality time with the family. Morgan said,

Once the convenience trade-off was equalized— YourGrocer makes local shopping just as convenient as using a supermarket—then other trade-offs, such as quality and supporting the community, became the differentiators. That's what sets us above the supermarkets. That's the real progress that people are able to make with us.

WHAT'S THE JTBD?

From the data Morgan has given us, I'd say that the struggle for progress is as follows:

Case Study: Morgan and YourGrocer

67

More about: My family having quality food, taking away the stress from grocery shopping, more family time, convenience

Less about: Grocery shopping online/supermarket /local shop, supporting the local community

Again, any kind of task or activity associated with grocery shopping is just a solution for a JTBD—it's not part of the JTBD itself. I know people who employ housemaids to keep the household fridge stocked with food and groceries. That entails no shopping at all—you pay someone else to take care of it. For those who can't afford or don't like that solution, grocery-delivery service is a nice alternative.

The progression of solutions in this case study helps us understand what customers do and don't value. In the beginning, parents were fine visiting multiple shops. They were willing to trade convenience for food quality. But when their family grew, saving time and reducing stress became more important to them. This is how we know that their desire to evolve, their

27024082R00022

Printed in Great Britain
by Amazon